WAVES IN MOTION

I0891794

LIGHT WAVES

by Julia Garstecki

PEBBLE
a capstone imprint

Pebble Emerge is published by Pebble, an imprint of Capstone.
1710 Roe Crest Drive
North Mankato, Minnesota 56003
www.capstonepub.com

Copyright © 2021 by Capstone. All rights reserved. No part of this publication may be reproduced in whole or in part, or stored in a retrieval system, or transmitted in any form or by any means, electronic, mechanical, photocopying, recording, or otherwise, without written permission of the publisher.

Library of Congress Cataloging-in-Publication Data is available on the Library of Congress website.
ISBN: 978-1-9771-2272-8 (library binding)
ISBN: 978-1-9771-2616-0 (paperback)
ISBN: 978-1-9771-2299-5 (eBook PDF)

Summary: Light waves are all around us. Learn all about light waves and how we use them every day.

Image Credits
Capstone Studio: Karon Dubke, 15, 20; Shutterstock: Africa Studio, Cover (wood), atsurkan, 17, Designua, 9, George Rudy, 19, Jochen Kost, 7, kosam, 8, Kuki Ladron de Guevara, 11, Mykola Mazuryk, 12, nadiya_sergey, 6, Nasky, 13, PR Image Factory, 16, Roman Yastrebinsky, Cover (lamp), WeAre, 5

Design Elements
Capstone; Shutterstock: Miloje, Nasky

Editorial Credits
Editor: Michelle Parkin; Designer: Ted Williams; Media Researcher: Jo Miller; Production Specialist: Laura Manthe

All internet sites appearing in back matter were available and accurate when this book was sent to press.

Printed in the United States 6257

TABLE OF CONTENTS

Words in **bold** are in the glossary.

WHAT ARE LIGHT WAVES?

Have you ever woken up in the middle of the night? You can't see well in the dark. What is that in the corner?

Turn on the lamp! It is only a pile of clothes. Light **waves** from the lamp light up your room. Light waves help us see. But how?

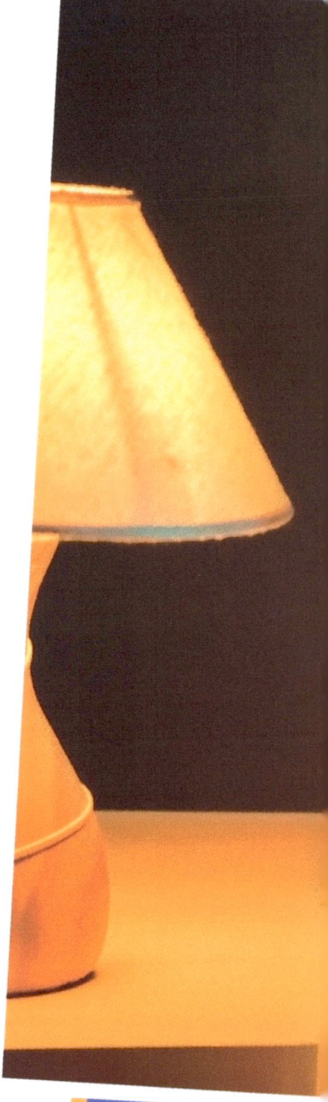

LIGHT WAVES FROM THE SUN

Most of the light we see comes from the sun. It travels here in waves. The sun is about 93 million miles away.

Light waves are fast. The waves can reach earth in 8 minutes and 20 seconds. Nothing is faster than the speed of light.

A COLORFUL WORLD

A tree's leaves are green. Its trunk is brown. We see these colors because of light waves. Light waves hit the tree. Some waves are soaked up, or **absorbed**.

Some light waves bounce off the tree.

These waves are **reflected** back to us.

Reflected waves are the colors we see.

REFRACTION

Light waves travel in a straight line. But some objects can bend the light. Put a pencil in a glass of water. The pencil looks like it is bending. But why?

This is called **refraction**. Light waves bend when they travel through the water. The pencil isn't bent. It only looks that way!

RAINBOW SCIENCE

Rainbows refract and reflect light. Look up at the sky after it rains. You may see a rainbow.

Light waves pass through water droplets in the air. These droplets bend the waves. This lets us see the rainbow's colors.

SHADOWS

Look behind you on a sunny day. Is something following you on the ground? If you jump, it jumps. If you run, it runs. It's your **shadow**!

Light waves can't pass through your body. You absorb some of the waves. The light waves you block make your shadow. Anything that blocks light has a shadow.

LIGHT INVENTIONS

We use light waves every day. We can't see some of the waves. Light waves in your wireless controller help you play a video game.

Some people need to wear eyeglasses. **Lenses** in the glasses bend the light. They help people see better.

HELPING WAVES

Light waves are all around us. They help us see colors. Some objects absorb light waves. Others reflect waves back to us.

Light waves help us use video games. They help us see in the dark. You can use a night light in your room. How do light waves help you?

CHANGE LIGHT WAVES

Find out how different objects can change light waves.

What You Need:

- a dark room
- flashlight
- empty glass
- glass of water
- wax paper
- piece of cardboard
- hand mirror

What You Need:

1. Find a small room, such as a bathroom.
2. Close the door. Make sure no light can come in from a window.
3. Place the empty glass, glass of water, wax paper, cardboard, and hand mirror on a table.
4. Turn off the light.
5. Shine your flashlight on each item.

Watch how the beam of light changes when it shines on each object. Does the light look the same? Why or why not?

GLOSSARY

absorb (ab-ZORB)—to soak up

lens (LENZ)—a piece of curved glass or plastic used in eyeglasses

rainbow (RAYN-boh)—arc of colors in the sky formed by sunlight shining through drops of water in the air

reflect (ree-FLEK)—to return light from an object

refraction (ri-FRAKT-shun)—the bending of light

shadow (SHAD-oh)—the dark shape when something blocks light

wave (WAYV)—the energy moving through air or water

READ MORE

Adler, David A. *Light Waves.* New York: Holiday House, 2018.

Briggs, Korwin. *The Invention Hunters Discover How Light Works.* New York: Little, Brown and Company, 2020.

Higgins, Melissa. *Light.* North Mankato, MN: Capstone, 2020.

INTERNET SITES

Light Facts

https://www.scienceforkidsclub.com/light.html

Science of Light

https://www.ducksters.com/science/light.php

INDEX